BOOKS EDITED BY JOHN L. CASTI AND ANDERS KARLQVIST:

daries and Barriers: On the Limits to Scientific Knowledge

MISSION TO AB

OTHER

Bour